INTERNATIONAL CENTRE FOR MECHANICAL SCIENCES

COURSES AND LECTURES - No. 85

JULIUSZ KULIKOWSKI

ACADEMY OF SCIENCES - WARSAW

ALGEBRAIC METHODS IN PATTERN RECOGNITION

**COURSE HELD AT THE DEPARTMENT
OF AUTOMATION AND INFORMATION
JULY 1971**

Springer-Verlag Wien GmbH 1971

© 1972 by Springer-Verlag Wien
Originally published by Springer-Verlag Wien-New York in 1972

ISBN 978-3-211-81128-3   ISBN 978-3-7091-2884-8 (eBook)
DOI 10.1007/978-3-7091-2884-8

P R E F A C E

During the last years electronic computation and data processing methods have reached a comparatively high level of maturity. More and more complicated forms of input data: numerical, alpha-numerical, textual can be handled automatically and processed by electronic computers. Last time a growing attention is paid to the picture processing. Graphical and pictorial forms of information play an important role anywhere a human individuum is linked on an informational system as its element. This is a consequence of the fact that a human visual tract is distinguished by its possibility to parallel information processing and by its flexibility to changing circumstances. That is why the problem of immediate visual information exchange between man and computer arises as a very actual one. The recognition of patterns by specialized automata or by electronic computers is a first step toward the solution of this problem. The pictures being of interest in many practical applications are not only typewritten or handwritten texts, but also graphics, electronic schemas, graphs, fingerprints, meteorogical charts, microphotos etc. Under certain assumptions they can be divided into less or more formalized classes to be discriminated automatically. It seems to us reasonable to consider the pic-

tures as some expressions of a "planar" language, sub-
jected to a certain number of morphological and syntactical rules. An identification of the sets of grammatical rules corresponding to the given classes of
patterns is a problem of interest if a pattern recognition algorithm is to be chosen. No universal "planar"
language seems to exist and no universal pattern recognition algorithm for a computer seems to be possible, as well. Nevertheless, the linguistic approach
turns out as a more effective one when composite patterns are dealt with. The other methods based on geometrical, statistical or functional models of patterns
recognition can be included as fragmentary ones in the
structural-linguistic analusis of a composite pattern.
A need of an universal meta-language for the description and theoretical investigation of "planar" languages that are used in particular situations arised.
In our opinion, this can be reached on the basis of a
general approach offered by a general theory of relations. This last can be specified in such a way that
any simple or composite picture is considered as a
realization of a generalized relation described on an
ordered family of sets of local input signal values.
This idea developed in the below given lectures. It
will be shown that a kind of Boolean algebra can be
defined on the set of possible relations. The well
known methods of Boolean functions minimization thus
can be used in order to minimize the length of an expression formally describing a pattern.

The considerations will be illustrated by numerical examples. However, they could not be considered as recommendations for the solutions of practical problems.

This is my duty and a great pleasure to express here my best thanks to Professor Luigi Sobrero and to Professor Angelo Marzollo for their initiative of including the lectures being given below to the program of CISM Summer School in June, 1971.

Udine, June 1971

1. Introductory remarks.

The recognition of patterns will be here consider ed as a cybernetical subdiscipline investigating the general principles of decision making in transmission systems with stationary information sources, under the assumption that the set of decisions is enumerable. A general model that will be here considered is given in Fig. 1. It consists, as usually, of an "information source", a "coder", a "transmission channel" containing a "noise source", a "decoder" and a "decision system". However, no special assumptions concerning the physical nature of the signals and the properties of the messages like this one con

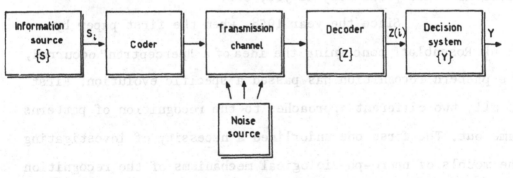

Fig. 1

cerning the finity of the set of messages or another concerning the statistical nature of the signals will be made. The problem of detection of the signals in presence of noise is a particular case of pattern recognition: it is supposed in this case, that the set of messages is finite and and the signals can be consid

ered as some realizations of stochastic processes. A problem of
typewritten text reading and interpretation is an example of
more complicated pattern recognition problem: the statistical
properties of the signals are of less importance in this case,
however, the difficulty is connected with the fact that the sets
of transmitted messages and the sets of decisions are infinite
now. A problem of identification of a speaking individual based
on an analysis of the sound of his voice is a particular case of
the recognition of a pattern, where the signals are stochastic
with a priori unknown statistical properties: the decision algo
rithm may be based on a sequential non-parametric statistical
technique. A problem of recognition of a geometrical form and dimensions of a rigid body by touch is a pattern problem having
no formal interpretation, as yet, etc.

Since the year 1958, when the first paper by
Frank Rosenblatt concerning the idea of a "perceptron" occurred,
the pattern recognition has passed a specific evolution. First
of all, two different approaches to the recognition of patterns
came out. The first one underlined a necessity of investigating
the models of neuro-physiological mechanisms of the recognition
processes in cerebral systems. The ideas of F. Rosenblatt played
an initiating role for this kind of investigations, which in
some sense can be also referred to the former concepts of Mc-
Culloch and Pitts of artificial neuron-sets. The other approach
is interested in investigations of new principles of pattern rec

ognition rather than in modelling the natural visual and aural tracts. This last concept is relative to the former ideas of optimum signal detection and signal parameters extraction. Nevertheless, the problem of pattern recognition is more general and difficult; it consists in finding out the algorithms of recognition and classification of type- and handwritten characters, graphical pictures, symptoms of diseases, microscopic images etc. The statistical formal models of signal reception are no more than a particular case of a general formulation of the pattern recognition problem.

Let us denote by S a message transmitted by the source and by $\{S\}$ a set of all possible transmitted messages, by Z – a signal observed at the output of the transmitted channel, by Y – a decision made by a pattern recognition device. The space of all possible signals observed at the input of the pattern recognition device will be denoted by $\{Z\}$, while $\{Y\}$ will denote a set of all possible decisions. The pattern recognition rule will be formally given in the form of a function

$$Y = h(Z) \qquad (1.1)$$

projecting $\{Z\}$ on $\{Y\}$ in an unique manner. It will be supposed that $\{Y\}$ is enumerable and that x_Y is its cardinal number.

Let Σ be an enumerable family of subsets $\{S\}_i \subset \{S\}$ and let x_Σ denote the cardinal number of Σ. It will be supposed that

$$x_\Sigma = x_Y \qquad (1.2)$$

and that there exists some one-to-one projection between $\sum$ and $\{Y\}$. Otherwise speaking, if $\{S\}_i$ is a "pattern", there exists a "decision" $Y_i \in \{Y\}$ corresponding to this pattern. The patterns will be here considered as some classes of messages rather than as single messages. For example, if a graphical representation of a given triangle on a plane is a message, then the pattern can be defined as a set of representation of all possible trian gles of different form and sizes. However, the graphical repre sentations transmitted through a "channel" (no matter what is the physical nature of this channel) are usually disturbed by the influence of some external processes. That is why pattern transmissions can be formally described as a projection of $\{S\}$ into $\{Z\}$, in general ambiguous and irreversive. Let $\{Z\}_i$ be a subset of signals at the input of the pattern recognition device corresponding to the pattern $\{S\}_i$. The pattern recognition prob lem will be called simple if

$$(1.3) \qquad\qquad \{Z\}_i \cap \{Z\}_j = \emptyset$$

for any $i = j$, where $\emptyset$ is an empty set, otherwise it will be cal led a complex one. The complex pattern recognition problems are typical in the case of stochastic disturbances or additive noise.

A general theoretical model of pattern recogni- tion, which can be considered as a "classical" one, was based on the following geometrical interpretation. Let us suppose, that $\{Z\}$ is a multidimensional Euclidean space E. It will be supposed that it is partitioned into a finite family of non-over lapping "cells" C_i bounded by a set of hypersurfaces of a given degree of regularity [11]. Then the "decision function"   given by (1.1) may be realized in the form

$$Y = Y_i \text{ if and only if } Z \in C_i, \quad i = 1,2,3\dots . \quad (1.4)$$

So as the geometrical model of pattern recognition is general enough, it seemed at the first stage that the main task of the pattern recognition theory consists in the investigation of methods of optimum space partitioning into the cells, according to the properties of the patterns. Even if it is so, it soon occurred that the problem is much more complicated in practice. The situation was well illustrated by one of the debaters at a Symposium in Soukhanovo (USSR) in 1967, who remarked that "it is impossible to construct in practice a surface separating... water from a sponge, even if the existence of such a surface is theoretically evident. The well known methods used in the pattern recognition theory, based on the statistical decision functions [6, 7], potential functions [1, 3], geometrical considerations [7, 8, 11] and many others, work well only in the case of relatively simple patterns. In the year 1962 a new idea by R. Narasimhan [24] has been published, which initiated the investigations of so called structural methods in pattern recognition. The structural approach consists in multi-level analysis of a picture. First of all the presence of some specific local features is detected on the basis of "classical"

methods. The relations between the local features are investigat
ed at the next level of picture analysis. Sometimes, the higher-
level relations between the relations of lower-degree are detec
ted at last. A kind of description of the picture in a formal
language instead of a simple decision is obtained as a result of
a structural approach, that is why this last is called sometimes
a linguistic approach. The local features can be considered as
the elements of a "vocabulary" of the language, while the higher-
degree relations form a sort of a "grammar" of the language.

      For example, it is not easy to represent in a geo
metrical form the set of all possible multidimensional binary-
component vectors representing all possible triangles projected
on a discrete retina, like this one:

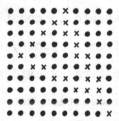

      The picture can be represented by a binary-component
vector in a 100-dimensional signal-space. This last is isomorphic
to a set of vertexes of a unity-cube plunged into a 100-dimension
al euclidean space. Nevertheless, a description of the set of vertexes corresponding to the images of all possible triangles, a-
part from their forms, dimensions and positions on the retina, is
hardly to be done. However, it is possible to obtain the tests for
the occurrence of the following local features: "a segment of a

straight line", "an angle between two segments", "an end-point
of a straight-line segment", and so on. Therefore, a "triangle"
can be defined as a simultaneous occurrence of the following
features: "three straight-line segments and three angles between
the segments". However, it is not enough: a constraint must be
put on the number of the free end-points of the segments: the
segments form a triangle if and only if this last number is
null. All this can be expressed in a more formal way. Let us
denote by A the fact that a straight-line segment has been detected on the retina, by B – the fact that two segments form an
angle, by C – the fact that a segment has a free end-point. An
index will be added to any symbol in order to discriminate between the independent facts. A formal description of a "triangle" will be given in a form of phrase:

$$D = A_1 \wedge A_2 \wedge A_3 B_1 \wedge B_2 \wedge B_3 \wedge \neg (C_1 \vee C_2 \vee C_3 \vee C_4 \vee C_5 \vee C_6).$$

A triangle is detected if and only if D is true
in logical sense.

There is a great difference between the decisions
made at the lower and at the higher levels. At the higher level
not only the set of all possible decisions, but also some relations between the decisions are a priori given. Otherwise
speaking, the higher-level decisions are not independent one
on each other. For example, if three straight-line segments
are detected, it is evident that no more than twelve angles be

tween them exist; if only three angles occur, it is evident that
the number of free end-points can be three or null, etc. Otherwise speaking, there is a set of relations between the potentia_l
ly possible higher-level decisions. As a consequence, the deci-
sion-making process is like a reasoning rather than like a simple
detection of the higher-level features occurrence.

Sometimes, the above mentioned relations can be
described using the well known graph-theory formalism [32] . Let
us consider the following example. The structural description of
the patterns of interest is based on the occurrence of two local
features, A and B. The set of all possible input signals can be
partitioned into the cells illustrated in Fig. 2. If the detec-

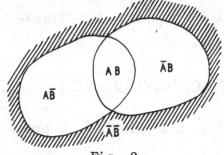

tion of the first feature can
give us an answer A or $\bar{A}$ alter-
natively, and the corresponding
results of the detection of the
second feature can be B or $\bar{B}$,

Fig. 2   then the alternative cells will

be AB, $\bar{A}B$, $A\bar{B}$ and $\bar{A}\bar{B}$. The implications of the first-level decisions on the second-level ones can be illustrated by a graph
shown in Fig. 3. This simple schema can be used in 14 different
pattern recognition problems:

  1. AB against $\bar{A}B$ or $A\bar{B}$ or $\bar{A}\bar{B}$,

  2. $\bar{A}B$ against AB or $A\bar{B}$ or $A\bar{B}$

  3. $A\bar{B}$ against AB or $\bar{A}B$ or $\bar{A}\bar{B}$

4. $\overline{A}\overline{B}$ against AB or $\overline{A}B$ or $A\overline{B}$,

5. AB or $\overline{A}B$ against $A\overline{B}$ against $\overline{A}\overline{B}$,

6. AB or $A\overline{B}$ against $\overline{A}B$ against $\overline{A}\overline{B}$,

7. AB or $\overline{A}B$ against $\overline{A}\overline{B}$ against $A\overline{B}$,

8. $\overline{A}\overline{B}$ or $A\overline{B}$ against AB against $\overline{A}B$,

9. $\overline{A}\overline{B}$ or $\overline{A}B$ against AB against $A\overline{B}$,

10. $A\overline{B}$ or $\overline{A}\overline{B}$ against AB against $\overline{A}B$,

11. AB or $\overline{A}B$ against $A\overline{B}$ or $\overline{A}\overline{B}$,

12. AB or $A\overline{B}$ against $\overline{A}B$ or $\overline{A}\overline{B}$,

13. $\overline{A}\overline{B}$ or $A\overline{B}$ against AB or $\overline{A}B$,

14. AB against $\overline{A}B$ against $A\overline{B}$ against $\overline{A}\overline{B}$.

(some of the formulations can be evidently simplified). The sec‐
ond—level decision sys‐
tem following the ar‐
rows on the graph can
reach its second—lev‐
el decision the first—
level decisions being
given. It is evident
that a strong relation

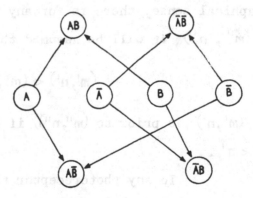

Fig. 3

between the logical structure of the higher—level decisions tech‐
nique and the internal structure of the patterns exists. A fam‐
ily of alternative patterns is no more a net of isolated ele‐
ments, it should be described using a more advanced mathematical
technique. On the other hand, the mathematical assumptions should

not be very strong if the theory is general. Therefore, the pat
tern description methods based on the set theory, topological
and algebraical properties are preferred by comparison with
these based on the probabilistic, geometrical or functional ones,
till the general recognition problem is of interest.

Some further assumptions concerning the set $\{Z\}$
of input signals will be necessary. There will be considered a
finite set of points on a plane forming a rectangular retina;
any ordered pair (m,n) of the integers $m \in <\overline{1,M}>, n \in <\overline{1,N}>$ will
describe an address of receptor, the photosensitive retina is
consisting of. Any subset of the photoreceptors will be supposed
to be linearly ordered, according to their addresses, in a lex-
icographical sense, there is for any pair of the addresses $(m',$
$n')$, (m'', n'') it will be assumed that

$$(m',n') < (m'',n'')$$

(read (m',n') is prior to (m'',n'')) if $m' < m''$ or if $m' < m''$
and $n' < n''$.

To any photoreceptor there will be assigned a
finite set

$$(1.5) \quad \{Z\}_{m,n} = U_{m,n} = \{0,1,...,k-1\}, \quad k = 2,3,4,...,$$

of possible local values of the input signals at the point (m,n)
The set of all possible input signals can be defined as a carte
sian product of the sets:

$$\{Z\} = \mathop{X}_{n=1}^{N} \mathop{X}_{m=1}^{M} \{X\}_{m,n} , \qquad (1.6)$$

any input signal

$$z = |z_{1,1}, z_{1,2}, \ldots, z_{M,N}| \in \{Z\} \qquad (1.7)$$

will then be represented by an NM-dimensional vector with the components belonging to a k-element finite set. The number k will represent the number of shadow levels, so as our considera tions will be mostly concerned to the monochromatic graphical pattern recognition. The "decision cells" corresponding to the patterns will be in general given in the form of subsets

$$C_i = \{Z\}_i \subset \{Z\} . \qquad (1.8)$$

The input signals $z \in C_i$ will be called the assumed realizations of the i-th pattern. It must be discriminated between the assumed and the real realizations of a pattern $\{S\}_i$; this last can in general belong to any decision cell C_j, j not obviously being identical with i . The pattern recognition is true if $j = i$, otherwise it is false. The real realization of the i-th pattern will be denoted by $z(i)$.

Our aim consists in choosing the decision cells C_i in a satisfactory if not optimum manner; it will be reached using the algebraical methods, at least at the higher decision-levels. The basic concept of our theory is this one of a generalized relation.

      If U' and U'' are some non-empty sets and $U' \times U''$ is their cartesian product, then any subset of this product

(1.9) $$R \subset U' \times U''$$

is called a relation between the elements $u \in U'$ and those ones $v \in U''$. For example if

$$U' = \{0,1\},$$
$$U'' = \{0,1\},$$

then

$$U' \times U'' = \{(0,0),(0,1),(1,0),(1,1)\},$$

and

$$R_1 = \{(0,0),(1,1)\},$$
$$R_2 = \{(0,1),(1,0)\}$$

will be some examples of relations (R_1 is called an identity relation and R_2-a negation). Let us remark that no special assumptions concerning the nature of the sets U' and U'' have been made. Therefore, it is possible to define the relations on the sets of points, on the families of sets, on the classes of func tions and on the sets of relations as well. It is then possible to build up a specific multilevel "construction" of the relations,

a possibility being of interest in the description of composite
patterns. However, for our purposes the assumption of a single
cartesian, as it has been made in (1.9) seems to be not necessary. Therefore, a generalized definition of a relation will be
introduced.

Let $<U_1, U_2, ..., U_N>$ be a linearly ordered family of
sets and let

$$U = \overset{N}{\underset{n=1}{\times}} U_n \tag{1.10}$$

be a cartesian product of the sets taken in the corresponding
order. Any subset

$$R \subset U \tag{1.11}$$

will be called a (generalized) relation between the elements
belonging to $U_1, U_2, ...,$ and U_n . A decision cell in the above
mentioned sense is a generalized relation between the signals
observed simultaneously by the photographers of a retina. However, our concept should be rather considered in a formal sense.
For example, if $U_1, U_2, ..., U_6$ are the sets of all natural numbers,
a certain relation R can be defined as a set of all series of
integers $<u_1, u_2, ., u_6>$ such that

$$v = \frac{u_1}{u_2} u_3 + \frac{u_4 - u_5}{u_6}$$

is an integer.

The relation R will be called empty if R is an
empty subset of U and will be called a trivial one if it is iden

tical with U. The empty relation will be denoted by θ. Any ordered set of values $<u_1, u_2, ..., u_N>$ belonging to $U_1, U_2, ..., U_N$ correspondingly, and such that

(1.12) $$u = <u_1, u_2, ..., u_N> \in R$$

will be called a realization of the relation R.

2. Algebra of relations.

Let us take into account a finite family of sets $U_1, U_2, ..., U_N$ and let R_1, R_2 be two arbitrary relations described on the cartesian product U defined by (1.10). So as the relations are defined as some subsets of U it is possible to combine the relations using the well known set-algebra operations. Therefore,

a) a disjunction of relations

(2.1) $$R = R_1 \cup R_2$$

can be defined as a relation, which is satisfied by all the realizations satisfying to at least one of the relations R_1, R_2 ;

b) a conjunction of relations

(2.2) $$R = R_1 \cap R_2$$

can be defined as a relation, which is satisfied by all the realizations satisfying to both the relations R_1 and R_2 ;

c) an asymmetrical difference of the relation

$$R = R_1 \div R_2 \qquad\qquad (2.3)$$

can be defined as a relation, which is satisfied by all the realizations satisfying to the relation R_1 and simultaneously not
satisfying to the relation R_2 , etc.

The following example will illustrate the above
concepts. Let $U_1, U_2, ..., U_9$ represent a set of states of a ret
ina-segment of a following form:

$$\langle U_1 \ U_2 \ U_3$$
$$U_4 \ U_5 \ U_6$$
$$U_7 \ U_8 \ U_9 \rangle$$

The following relations will be introduced:

a) "a horizontal line in the upper-half of the segment"
$$R_a = \{\langle 111\ 000\ 000\rangle, \langle 000\ 111\ 000\rangle\},$$

b) "a horizontal line in the down-half of the segment"
$$R_b = \{\langle 000\ 111\ 000\rangle, \langle 000\ 000\ 111\rangle\};$$

then the following new relations can be defined:

c) "a horizontal line through the segment"
$$R_c = R_a \cup R_b = \{\langle 111\ 000\ 000\rangle, \langle 000\ 111\ 000\rangle, \langle 000\ 000\ 111\rangle\},$$

d) "a horizontal line going through the centre of the segment"

$$R_d = R_a \cap R_b = \{<000\ 111\ 000>\},$$

e) "a horizontal line at the top of the segment"

$$R_e = R_a \dot{-} R_b = \{<111\ 000\ 000>\},$$

etc. Generally speaking, it is possible to define 2^{k^N} different relations on a cartesian product of N sets, each containing k elements. However, it is not necessary to operate with all these relations for an effective recognition of patterns.

The relation R_a will be called a subrelation of the relation R_b if

(2.4) $$R_a \cup R_b = R_b \quad \text{and} \quad R_a \cap R_b = R_a.$$

The relations R_a, R_b will be called mutually disjoin if

(2.5) $$R_a \cap R_b = \theta.$$

Till now the algebraic operations on the relations described on the same family of sets were considered. However, it seems desirable to make something similar in the case if the families are not the same, for example, if the relations are described on different segments of the retina. Let us consider the family of sets $\{U_1, U_2, ..., U_N\}$. All ahead, if a subfamily $\{U_{\nu_1}, ... , U_{\nu_p}\}$ of sets is considered, it will be supposed that the linear order introduced in the full family of sets is conserved inside the subfamily. Let us take two subfamilies of sets into account: $\{U_{\nu_1}, .., U_{\nu_p}\}$ and $\{U_{\mu_1}, .., U_{\mu_q}\}$. Both they are linearly ordered in the sense, that

$$\nu_1 < \nu_2 < \ldots < \nu_p < N$$

and

$$\mu_1 < \mu_2 < \ldots < \mu_q < N$$

if $U_1, U_2, \ldots, U_N$ are ordered accordingly to their indices. Conse-
quently, the order will be supposed to be conserved if a disjunction, a conjunction or a difference of the subfamilies is
considered. The two above-mentioned linearly ordered subfamilies
of sets will be briefly denoted by $<U>'$ and $<U>''$, correspondingly. Let R' and R'' be some relations described on $<U>'$ and $<U>''$.

   Let $<U>''' \subset <U>'$ be a subfamily of sets, if $u' \in R'$
is a sequence of the elements satisfying to the relation R',
then $u''' \subset u'$ will denote a subsequence consisting of the elements of u' belonging to the sets belonging to the subfamily $<U>'''$
The set of all the subsequences u''' satisfying to this criterion will be denoted by R'''. It is evident, that R''' can be
considered as a new relation described on $<U>'''$, depending
in some sense on the relation R'. The relation R''' will be
called a projection of R' into the subfamily $<U>'''$:

$$R''' = R_{<U>'''} . \qquad (2.6)$$

   The following example will illustrate the last
idea. If

$$R_c = \{<111\ 000\ 000>, <000\ 111\ 000>, <000\ 000\ 111>\}$$

is a relation on the family $<U_1, U_2, ., U_9>$ of binary sets and $<U_2,$ $U_3, U_5, U_6, U_8, U_9>$ is a subfamily of sets, then

$$R = \{<11\ 00\ 00>, <00\ 11\ 00>, <00\ 11\ 00>\}$$

can be considered as a projection of the relation R_c ("a horizontal line through the 3×3 segment") into the 3×2 subsegment having the form

$$
\begin{aligned}
&\cdot <U_2\ U_3 \\
&\cdot\ \ U_5\ U_6 \\
&\cdot\ \ U_8\ U_9> .
\end{aligned}
$$

It is clear, that R will describe "a horizontal line through the subsegment".

Let $<U>'$ be a subfamily of a family $<U>$ of sets and let R' and R be some relations described on $<U>'$ and on $<U>$ correspondingly. We will take into account all the realizations of the relation R such that their projections on $<U>$ satisfy to the relation R'. The set R'' of realizations satisfying to this condition is a subrelation

(2.7) $R'' \subset R$

and the projection of R'' into $<U>'$ is a subrelation

(2.8) $R''_{<U>'} = R$

R'' will be called a conditional relation R for given R' and will be denoted by

(2.9) $R'' = R(R') .$

Let us consider the following example. There are given the former families of set

$$\langle U \rangle = \langle U_1, U_2, U_3, U_4, U_5, U_6, U_7, U_8, U_9 \rangle$$

$$\langle U' \rangle = \langle U_2, U_3, U_5, U_6, U_8, U_9 \rangle \subset \langle U \rangle .$$

The following relations are defined:

a) "a horizontal line through the segment"

$$R = \left\{ \langle 111\ 000\ 000 \rangle, \langle 000\ 111\ 000 \rangle, \langle 000\ 000\ 111 \rangle \right\},$$

b) "a line going through the element N° 6"

$$R' = \left\{ \langle 01\ 01\ 01 \rangle, \langle 10\ 01\ 00 \rangle, \langle 00\ 11\ 00 \rangle, \langle 00\ 01\ 10 \rangle \right\}.$$

The conditional relation "a line going through the full segment given a line going through the element N 6" will have the form

$$R'' = R(R') = \left\{ \langle 000\ 111\ 000 \rangle \right\}.$$

The concept of conditional relation can be easily generalized. Let $\langle U' \rangle$ and $\langle U'' \rangle$ be two families of sets and let us denote by

$$\langle U''' \rangle = \langle U' \rangle \cap \langle U'' \rangle \qquad\qquad (2.10)$$

$$\langle U^{iv} \rangle = \langle U' \rangle \cup \langle U'' \rangle \qquad\qquad (2.11)$$

the conjunction and the disjunction of the families; it will be supposed that $\langle U''' \rangle$ is a non-empty family of sets. Let us suppose that R' and R'' are two relations described on $\langle U' \rangle$ and $\langle U'' \rangle$ correspondingly. We will take into account the projections $R'_{\langle U \rangle} \ldots$

and $R''_{<U>} \cdots$. Let us take their conjunction:

(2.12) $$R^v = R'_{<U>} \cdots \cap R''_{<U>} \cdots \,.$$

The relation R^v consists of all the subsequences
of the elements belonging to the sets of $<U'''>$ such that any sub
sequence is a common part of two sequences satisfying to the
relation R' and R'' correspondingly. The two above-mentioned
sequences can be joined into a sequence belonging to the family
$<U>^{iv}$ of sets. The set of all joined sequences will be denoted
by R ; it can be considered as a new relation described on the
family $<U>^{iv}$, connected in some sense with the relations R' , R'' .
It will be called a convolution of the relations

(2.13) $$R = R' * R'' \,.$$

Now it is clear, that the conditional relation
$R'(R'')$ is a particular case of the convolution $R' * R''$ obtained
in the case if $<U''> \subset <U'>$. The following example will illustrate
the idea of convolution.

Let us consider a segment of the retina having
the following form

$$<U_1 \, U_2 \, U_3$$
$$U_4 \, U_5 \, U_6 \, U_7$$
$$U_8 \, U_9 \, U_{10} \, U_{11}$$
$$U_{12} \, U_{13} \, U_{14}> \,.$$

The two families of sets will be choosen in the following way:

$$\langle U \rangle' = \langle U_1, U_2, U_3, U_4, U_5, U_6, U_8, U_9, U_{10} \rangle,$$

$$\langle U \rangle'' = \langle U_5, U_6, U_7, U_9, U_{10}, U_{11}, U_{12}, U_{13}, U_{14} \rangle.$$

The conjunction of the families has the form

$$\langle U \rangle''' = \langle U \rangle' \cap \langle U \rangle'' = \langle U_5, U_6, U_9, U_{10} \rangle.$$

The two relations called "a line through the centre of the seg-
ment" will be defined on $\langle U \rangle'$ and $\langle U \rangle''$ correspondingly:

$$R' = \{\langle 000\ 111\ 000 \rangle, \langle 010\ 010\ 010 \rangle, \langle 100\ 010\ 001 \rangle, \langle 001\ 010\ 100 \rangle\},$$

$$R'' = \{\langle 000\ 111\ 000 \rangle, \langle 010\ 010\ 010 \rangle, \langle 100\ 010\ 001 \rangle, \langle 001\ 010\ 100 \rangle\}.$$

It should be remarked, that R' and R'' being defined on different segments are different relations, although the
formulae are similar. The convolution of the relations has the
form

$$R = R' * R'' = \{\langle 100\ 0100\ 0010\ 001 \rangle\}$$

and describes a straight line going through the elements № 1,
5, 10 and 14 of the retina.

It is also evident that in the case if $\langle U \rangle' = \langle U \rangle''$,

$$R' * R'' = R' \cap R'', \qquad (2.14)$$

therefore, the convolution of relations can be regarded as a gen-

eralization of the conjunction of the relations. A question a-
rises if it is also possible to define a new operation on the
relations. Let R' and R'' be some relations described on $\langle U \rangle'$
and $\langle U \rangle''$ correspondingly. We will take into account the follow
ing subfamilies of the sets:

$$(2.15) \qquad\qquad \langle U \rangle''' = \langle U \rangle' \div \langle U \rangle'',$$

$$(2.16) \qquad\qquad \langle U \rangle^{iv} = \langle U \rangle'' \div \langle U \rangle'.$$

Let us denote by R''' and R^{iv} the cartesian products
of the sets belonging to the subfamilies $\langle U \rangle'''$ and $\langle U \rangle^{iv}$ correspondingly. We will take into account the set of all sequences of ele-
ments belonging to the family $\langle U \rangle' \cup \langle U \rangle''$ of sets, obtained by
one of the following manners:

1° a realization of R' is joined together with a realization of R^{iv};

2° a realization of R'' is joined together with a realization
of R''.

The disjunction of the so obtained sets of se-
quences can be considered as a new relation R described on $\langle U \rangle'$
$\cup \langle U \rangle''$; the relation will be called a cartesian product of the
relations R' and R'' :

$$(2.17) \qquad\qquad R = R' \times R''.$$

Let us consider the following example. There is
given a family of binary sets and the relations R' and R'' like
those considered in the example illustrating the idea of convolu
tion. The differences of the set families are as follows:

$$<U>''' = <U>' \div <U>'' = <U_1, U_2, U_3, U_4, U_8> \, ,$$

$$<U>^{iv} = <U>'' \div <U>' = <U_7, U_{10}, U_{12}, U_{13}, U_{14}> \, .$$

It is clear that the number of all possible elements of the cartesian product of the sets belonging to $<U>'''$
is 2^5, this is also the number of the elements belonging to the
cartesian product of sets belonging to $<U>^{iv}$. Joining all possible realization of R' with those ones of R^{iv} as well as the realization of R''' with those ones of R'' we obtain:

$$R'xR'' = \{<000\ 111.\ 000.\ \dots>, <010\ 010.\ 010.\ \dots>,$$

$$<100\ 010.\ 001.\ \dots>, <001\ 010.\ 100.\ \dots>,$$

$$<\dots\ .000\ .111\ 000>, <\dots\ .010\ .010\ 010>,$$

$$<\dots\ .100\ .010\ 001, \ \dots\ .001\ .010\ 100>\}$$

where any possible combination of zeros and unities can be put
into the places denoted by points. Therefore, the cartesian product $R' \times R'''$ consists here of $8 \cdot 2^5 = 256$ realizations, the to-

tal number of all possible binary combinations described on the disjunction of sets $\langle U \rangle' \cup \langle U \rangle''$ being equal here $2^{14} = 18\,384$.

The convolution of relations can be defined as a maximum set of realizations satisfying to both R' and R'' , while the cartesian product can be considered as a minimum set of the realizations satisfying at least to one of the component realizations R' even/or R'' .

It is evident that in general

$$(2.18) \qquad\qquad R' * R'' \subset R' \times R''$$

and that the cartesian product of relations is identical with the disjunction of relations if $\langle U \rangle' = \langle U \rangle''$ excepting the situation if one of the components R' or R'' is an empty relation; the car tesian product is not defined in this last case. However, in order to make the analogy between the disjunction and the carte sian product of relations full it is possible to postulate that

$$(2.19) \qquad\qquad R \times \theta = \theta \times R = R.$$

On the other hand, the convolution of relations is empty if the factors R', R'' do not "overlap"; it does mean that some of the relations are empty, even/ or no pairs of their realizations have common elements on the non-empty conjunction of their set families. The relations will be called excluding each other in this case.

Both convolution and cartesian product of the

relations, as a result of preservation of the linear order in
the families of sets, are commutative. Let us denote by $Q_{<U>}$ the
trivial relation described on some linearly ordered family of
sets $<U>$. Any relation R described on a subfamily $<U'> \subset <U>$ can be
easily extended on the full family $<U>$ if taking the relation
$Q_{<U>''} \times R$, where $<U''> = <U> \div <U'>$ instead of R . Therefore, a comple-
mentary relation $\bar{R}$ to the given R on the family $<U'>$ can be defined as

$$\bar{R} = Q_{<U'>} \div R, \qquad (2.20)$$

but this is equivalent to the relation

$$Q_{<U>''} \times (Q_{<U'>} \div R) = Q_{<U>} \div (Q_{<U>''} \times R) \equiv Q_{<U>} \div R . \qquad (2.21)$$

Now, let us remark that if we denote by $A(R)$
the phrase "there is given a realization satisfying to the relation R'', then, according to the definitions,

$$A(R' \times R'') \iff A(R') \lor A(R''), \qquad (2.22)$$

$$A(R' * R'') \iff A(R') \land A(R''). \qquad (2.23)$$

Therefore, it is clear that the convolution and
cartesian product of the relations will have formal properties
similar to those of the logical conjunction and alternative. In
particular, the de Morgan's identities hold:

$$\overline{R' \times R''} = \bar{R}' * \bar{R}'', \qquad (2.24)$$

(2.25) $$\overline{R' * R''} = \overline{R'} \times \overline{R''}.$$

It can be also proved the associativity law for both operations

(2.26a) $\quad R' \times (R'' \times R''') = (R' \times R'') \times R''' = R' \times R'' \times R''',$

(2.26b) $\quad R' * (R'' * R''') = (R' * R'') * R''' = R' * R'' * R''',$

as well as the distributivity law of any operation with respect
to each other:

(2.27a) $\qquad R' \times (R'' * R''') = (R' \times R'') * (R' \times R'''),$

(2.27b) $\qquad R' * (R'' \times R''') = (R' * R'') \times (R' * R''').$

Otherwise speaking, the ordered six elements $<<U>$, $Q_{<U>}, \theta, *, \times, \bar{\ }, >$ form a Boolean algebra with the empty relation
as a null and the trivial relation $Q_{<U>}$ as an unity element.

This last statement may be very important in prac
tice. It means that the well known methods of logical functions
minimization and constructive methods of logical systems design
can be applied for the design of picture interpretation and pat
tern recognition algorithms.

Although the main concepts have been illustrated
in the families of binary sets, the algebra of relations does
not put any constraints on the nature of the sets. For example,
let us take into account a linearly ordered set $<X_1, X_2, \ldots, X_N>$
of random variables. Let us define a set of relations $R_2, R_3, \ldots, R_N$
as follows: the relation R_n consists of all sequences $<X_{i_1}, X_{i_2}, \ldots$

$...,X_{i_n}>$ of the random variables statistically dependent in the set. A question arises of what are the relations (of higher or der) between the relations R_n . First of all, a relation P of "overlapping" can be defined as a sequence of the relations $<R_{n_1},$ $,R_{n_2},...,R_{n_k}>$ such that there exists at least one random variable X_i satisfying to all the relations of this sequence. However, it does not mean that there is a statistical dependence between the rest of the arguments of the relations $R_{n_1}, R_{n_2},..., R_{n_k},$ because it can be very easily proved that the statistical dependence between the random variables is not a transitive property. Nevertheless, if some random variables $X_{i_1}, X_{i_2},...,X_{i_n},$ are pairwise statistically dependent any one on each other, they are statistically dependent in common. Therefore, there exists a kind of relation between the relation R_n different from the P-relation: a T-relation can be defined as a relation consisting of all sequences of the relations $<R_{n_1}, R_{n_2},...,R_{n_m}>$ such that R_{n_m} is performed if all the preceding relations $R_{n_1}, R_{n_2},..,R_{n_{m-1}}$ are performed. A question then arises of what is the "highest order" relation between the relations P and T . For example, they are overlapping in the former sense. Investigations of this kind of relations between the random variables may be interesting from the point of view of an analysis of an informational system structure. However, this is not a problem of pattern recognition.

3. Structural approach to the recognition patterns.

An automatic classification of composite patterns consists of several levels of picture processing. The main levels are the following:
  a) rowing, framing, scaling, lightening adjustment etc.;
  b) filtering noise, correction of defects, narrowing lines, showing off contours etc.;
  c) detection of local features;
  d) higher level features detection and interpretation.

The above mentioned picture processing levels should not be considered as strictly independent on each other. Sometimes the higher-level decisions should imply backward, on the decisions of the ·lower-level.

The decision algorithms of the lower levels up to the detection of local features can be based on the "classical" recognition methods. The deterministic, statistical and learning algorithms can be used at this stage of picture processing. However, a problem usually arises of choosing a set of basic features forming a "vocabulary" of a language for the pattern description. The problem is usually solved by intuition; some authors tried to solve the problem using more advanced methods [6, 13] . An optimizing procedure is evidently possible if a criterion of optimization is defined. However, choosing the criterion is a serious problem itself. The pattern recognition problem can

be considered as a problem of reduction of the information redundance. For example, if the retina consists of N elements char
acterized by two possible states ("black-white") and the set of
final decisions contains possible elements (patterns to be discriminated), the reduction coefficient is equal

$$c = \frac{N}{log_2 x} \qquad (3.1)$$

and in a lot of practical cases is much larger than unity. The
process of informational redundance reduction is distributed
over the picture processing levels. However, a strong redundance
reduction at a single level leads to serious practical difficul
ties. If γ is a maximum reduction coefficient for a single lev
el, the number d of the necessary picture processing levels may
be obtained from the inequality

$$d\,log\gamma \geq log\,c . \qquad (3.2)$$

The necessary number x' of picture classes, which
are to be discriminated at the first level of picture processing,
may be obtained from the formula

$$log_2 x' = \frac{N}{\gamma} , \qquad (3.3)$$

and the problem of choosing the set of basic features becomes
equivalent to the one of optimum classification of the original
class of pictures into x' classes of the first discrimination
level. However, the so obtained x' number should not be confused

with the number of necessary local features to be detected, so
as the position (address) of the local feature occurrence carries
some additional information about the recognized pattern. For
example, if the local features are detected on a subsegment of
the dimensions $M' \times N'$, where $M' < M$, $N' < N$ and M, N are the dimensions of a rectangular retina, and if the "window" is shifted step
by step having b possible positions on the retina, the number of
possible results of the first level picture analysis can be coded using b symbols from an alphabet containing the number of elements equal to the number of all possible local decisions. Otherwise speaking, the necessary number of local features will be
$(x')^{1/b}$. Now, it is clear that it is possible, in general,
to reach the same result using a large set of local features of
more complicated form, described on the subsegments of large
dimensions, or a very limited set of local features of simple
form described on little subsegments.

A distinction between the deterministic detection
procedures for the local image feature and the statistical ones
is quite relative: the deterministic procedures can be obtained
from the statistical ones in asymptotical situations, when the
distribution variances decrease. In particular, the origin of
the well known feature detection methods based on the distance
measure between the local signal and the standards lies in the
statistical maximum likelihood technique. The following distance
measures are commonly used in recognition problems. Let us denote

by

$$z^* = <z_1^*, z_2^*, \ldots, z_n^*> \qquad (3.4)$$

a standard signal representing the local feature, which is to
be detected, and

$$z = <z_1, z_2, \ldots, z_n> \qquad (3.5)$$

is the local subrealization of a real input signal, which is to
be processed. The most popular distance measure is this one proposed by R.W. Hamming:

$$\varrho(z^*, z) = \sum_{\nu=1}^{N} (x_\nu^* - x_\nu). \qquad (3.6)$$

An euclidean distance measure is rather preferable
in the case of multi-stage components vectors:

$$\varrho(z^*, z) = \left[\sum_{\nu=1}^{n} (z_\nu^* - z)^2 \right]^{1/2}; \qquad (3.7)$$

in the case of the binary-components vectors the Hamming's and
the euclidean distances are equivalent.

Another distance measure was used by M.A. Aizerman, E. M. Braverman and L.I. Rosoneer [1, 3]. If $\{\psi_\nu(x)\}$ is a
set of scalar functions described on the set of vector arguments, a potential function can be defined as

$$K(z^*, z) = \sum_{\nu=1}^{N} \lambda_\nu^2 \, \psi_\nu(x^*) \psi_\nu(x), \qquad (3.8)$$

where λ_ν^2 are some real coefficients. Then, the distance measure
is defined as

(3.9) $\varrho(z^*,z) = \left[K(z^*,z^*)+K(z,z)-2K(z^*,z)\right]^{1/2}$.

This distance measure obtains an euclidean form in a particular case, if we suppose that

(3.10) $K(z^*,z) = (z^*,z) = \sum_{v=1} z_v^* z_v$.

If z^*, z are binary-compbnent vectors and

(3.11) $w(x) = \sum_{v=1}^{N} |z_v|$

is the weight of a vector (the number of unity-components), then it can be defined a

(3.12) $g(z^*,z) = \dfrac{(z^*,z)}{w(z^*)+w(z)-(z^*,z)}$

and a distance measure (proposed by J. Rogers and T. Tanimoto [31]) takes the form:

(3.13) $\varrho(z^*,z) = -\log_2 g(z^*,z)$.

The main distress of the detection procedures based on the distance measures consists in the fact that the decisions are not invariant with respect to the so called optical transformations of the pictures. The optical transformation of a picture given in the form of a vector with real components is described by the formula

(3.14) $T(z) = <az_1 + b,\ldots,az_n + b>,$

where a , b are some positive coefficients. The invariance of a detection procedure with respect to the optical transformations can be illustrated in the following example. The two pictures describing the same pattern (a letter "L") should be classified as belonging to the same class, in spite of the fact that they are represented by the vectors of different lengths:

```
0  0  0  0  0        1  1  1  1  1
0  1  0  0  0        1  3  1  1  1
0  1  0  0  0        1  3  1  1  1
0  1  0  0  0        1  3  1  1  1
0  1  1  1  0        1  3  3  3  1
0  0  0  0  0        1  1  1  1  1
```

Here, $a = 2$ and $b = 1$.

This difficulty can be avoided if a coincidence measure is used instead of the distance:

$$r(z^*, z) = \frac{\sum\limits_{\nu=1}^{N}(z_\nu^* - \bar{z}_\nu^*)(z_\nu - \bar{z}_\nu)}{\sqrt{\sum\limits_{\nu=1}^{N}(z_\nu^* - \bar{z}_\nu^*)^2 \sum\limits_{\mu=1}^{N}(z_\mu - \bar{z}_\mu)^2}}, \qquad (3.15)$$

where

$$\bar{z} = \frac{1}{N}\sum_{\nu=1}^{N} z_\nu, \qquad (3.16)$$

the $\bar{z}^*$ having a similar sense.

A lot of other examples of coincidence measures is given in the paper by G.N. Zhitkov in [35] .

If the distance measure or a correlation method of local detection is used, the problem arises of optimum choosing a set of standard signals z^* ; a lot of papers have been devoted to this problem (2, 4, 5, 7, 8, 10, 12); they will not be discussed here in more detail as they are not immediately connected with the algebraical methods.

The detection algorithm based on the correlation coefficient (3.15) becomes inconvenient in case the lowest levels of picture processing are to be realized by a logical set. Therefore, an approximation of the correlation algorithm by a set of logical operation is of interest.

Let us define a local pattern as "the end of a line-segment". This can be represented by the patterns having, for example' the following realizations:

```
1 0 0 0 0   0 0 0 1 0   0 0 0 0 0   0 0 0 0 0   0 0 0 0 0
1 0 0 0 0   0 0 1 1 0   1 0 0 0 0   0 0 0 0 0   0 0 1 0 0
0 1 0 0 0   0 0 1 0 0   1 1 1 0 0   1 1 1 0 0   0 0 1 0 0
0 0 1 0 0   0 0 0 0 0   0 0 0 0 0   0 0 1 0 0   0 0 0 1 0
0 0 0 0 0   0 0 0 0 0   0 0 0 0 0   0 0 0 0 0   0 0 1 0 0
```

etc. A test for this kind of local pattern can be described in the relation language as a relation satisfied by the 9-components vectors $z(m,n)$ of the form

$$z_{m-1,n-1} \quad z_{m-1,n} \quad z_{m-1,n+1}$$

$$z_{m-1,n-1} \quad z_{m,n} \quad z_{m,n+1}$$

$$z_{m+1,n-1} \quad z_{m+1,n} \quad z_{m+1,n+1} \ .$$

The relation $R_1(m,n)$ called "the end of a line-segment at the point (m,n) is given by the formula:

$$R_1(m,n) = \bigcup_{i=1}^{4} R_{11}^{(i)}(m,n) * R_{12}^{(i)}(m,n), \qquad (3.17)$$

where

$$R_{11}^{(1)}(m,n) = \left\{ \begin{matrix} <0 & 1 & . \\ 0 & 1> & . \\ . & . & . \end{matrix} , \begin{matrix} <1 & 0 & . \\ 0 & 1> & . \\ . & . & . \end{matrix} , \begin{matrix} <1 & 1 & . \\ 0 & 1> & . \\ . & . & . \end{matrix} , \begin{matrix} <1 & 0 & . \\ 1 & 1> & . \\ . & . & . \end{matrix} \right\} \qquad (3.18)$$

$$R_{12}^{(1)}(m,n) = \left\{ \begin{matrix} . & . & <0 \\ 0 & 1 & 0 \\ 0 & 0 & 0> \end{matrix} \right\}, \qquad (3.19)$$

(the point . indicates that the relation is not defined for the given component) and the relations $R_{11}^{(2)}, R_{12}^{(2)}, R_{11}^{(3)}, R_{12}^{(3)}, R_{11}^{(4)}, R_{12}^{(4)}$ can be obtained from the $R_{11}^{(1)}$ and $R_{12}^{(1)}$ by the corresponding counter-clock rotations of the square segment on 90, 180, and 270 degrees.

Applying a test for the $R_1(m,n)$ relation to the picture shown in the above-given example we shall obtain correspondingly:

```
. . . . .     . . . . .     . . . . .     . . . . .     . . . . .
. 0 0 0 .     . 0 0 0 .     . 0 0 0 .     . 0 0 0 .     . 0 1 0 .
. 0 0 0 .     . 0 1 0 .     . 0 1 0 .     . 0 0 0 .     . 0 0 0 .
. 0 1 0 .     . 0 0 0 .     . 0 0 0 .     . 0 1 0 .     . 0 0 0 .
. . . . .     . . . . .     . . . . .     . . . . .     . . . . .
```

the unities showing the addresses of the $R_1(m,n)$ relations being satisfied.

However, usually a picture is not given in such a distinct form, as supposed in the example. The lines are rather diffused like in the following examples:

```
0 1 1 1 0 0 0 0 0 0        0 0 0 0 0 0 0 0 0 0 0
0 0 1 1 1 1 0 0 0 0        0 0 0 0 0 0 0 0 0 0 0
0 0 0 1 1 1 0 0 0 0        0 0 0 1 1 1 1 1 0 0
0 0 0 1 1 1 1 0 0 0        1 1 1 1 1 1 1 1 0 0 0
0 0 0 0 1 1 1 0 0 0        1 1 1 0 0 1 1 1 0 0
0 0 0 0 1 1 1 1 0 0        1 1 0 0 0 0 1 1 1 0
0 0 0 0 0 1 1 0 0 0        1 1 0 0 0 0 1 1 0 0
0 0 0 0 0 0 0 0 0 0        1 0 0 0 0 0 0 0 0 0 0
0 0 0 0 0 0 0 0 0 0        1 0 0 0 0 0 0 0 0 0 0
0 0 0 0 0 0 0 0 0 0        0 0 0 0 0 0 0 0 0 0 0
```

and a "skeleton" of the picture must be shown off before the line-end is detected.

Let us assign to any component $x_{m,n}$ its "environment" given by the vector

$$\zeta(m,n) = <z_{m-1,n-1}, z_{m,n-1}, z_{m+1,n-1}, z_{m+1,n}, z_{m+1,n+1},$$

(3.20)

$$z_{m,n+1}, z_{m+1,n-1}, z_{m-1,n}> .$$

The weight of a vector z will be denoted by $w(z)$. The procedure of showing off the "skeleton" of a fuzzy picture will be based on the following transformation:

$$(3.21) \quad z_{m,n} := \begin{cases} 0 & \text{if } z_{m,n} = 0 \text{ or } z_{m,n} = 1 \text{ and } 2 < w(\zeta_{m,n}) \leqslant \theta, \\ 1 & \text{otherwise} \end{cases}$$

(the symbol: = should be read as "becomes equal"). Applying the formula (3.21) to the above given examples we obtain the following results:

```
.  .  .  .  .  .  .  .  .  .            .  .  .  .  .  .  .  .  .  .
.  0  0  1  1  0  0  0  0  .          .  0  0  0  0  0  0  0  0  .
.  0  0  1  1  1  0  0  0  .          .  0  0  0  0  0  0  0  0  .
.  0  0  0  1  1  0  0  0  .          .  1  1  1  1  1  1  0  0  .
.  0  0  0  1  1  1  0  0  .          .  1  0  0  0  0  1  0  0  .
.  0  0  0  0  1  1  0  0  .          .  1  0  0  0  0  1  1  0  .
.  0  0  0  0  0  0  0  0  .          .  0  0  0  0  0  0  0  0  .
.  0  0  0  0  0  0  0  0  .          .  0  0  0  0  0  0  0  0  .
.  0  0  0  0  0  0  0  0  .          .  0  0  0  0  0  0  0  0  .
.  .  .  .  .  .  .  .  .  .            .  .  .  .  .  .  .  .  .  .
```

The procedure can be repeated several times, if necessary. Applied once more it gives, correspondingly:

```
.  .  .  .  .  .  .  .  .  .  .            .  .  .  .  .  .  .  .  .  .  .
.  .  0  0  1  0  0  0  .  .            .  .  0  0  0  0  0  0  .  .
.  .  0  0  1  1  0  0  .  .            .  .  1  1  1  1  1  0  .  .
.  0  0  0  1  0  0  .                  .  0  0  0  0  0  1  0  .
.  .  0  0  0  0  0  0  .  .            .  .  0  0  0  0  1  0  .  .
.  .  0  0  0  0  0  0  .  .            .  .  0  0  0  0  1  1  .  .
.  0  0  0  0  0  0  .                  .  0  0  0  0  0  0  .
.  .  0  0  0  0  0  0  .  .            .  .  0  0  0  0  0  .  .
.  .  0  0  0  0  0  .  .              .  .  0  0  0  0  0  .  .
.  .  .  .  .  .  .  .  .              .  .  .  .  .  .  .  .  .
.  .  .  .  .  .  .  .  .              .  .  .  .  .  .  .  .  .
```

and both pictures are ready for the application of the algorithms proving the distinct local relations.

The situation gets more complicated if the picture is distorted by noise, as illustrated by the following examples:

original picture   fuzzy picture

```
0 0 0 1 1 1 0 0 0 0        0 0 0 1 1 0 0 0 0 0
0 0 0 1 1 1 0 0 0 0        0 0 0 1 1 1 1 0 0 0
0 0 0 1 1 1 0 0 0 0        0 0 0 0 1 1 0 1 0 0
0 0 1 1 1 0 0 0 0 0        0 0 1 1 1 0 0 0 0 0
0 0 1 1 1 0 0 0 0 0        0 0 1 1 0 0 0 0 1 0
0 0 1 1 1 0 0 0 0 0        0 0 1 0 1 0 0 0 0 0
0 0 0 1 1 0 0 0 0 0        0 0 0 1 1 0 0 0 0 0
0 0 0 0 0 0 0 0 0 0        0 0 0 0 0 0 0 0 1 0
0 0 0 0 0 0 0 0 0 0        0 0 1 0 0 0 0 0 0 0
0 0 0 0 0 0 0 0 0 0        0 0 0 0 0 0 0 0 0 0 .
```

Applying the former algorithm based on the "rust" technique we obtain the result

```
0 0 0 1 0 0 0 0
0 0 0 1 0 0 1 0
0 0 0 0 0 0 0 0
0 0 1 0 0 0 0 1
0 0 0 1 0 0 0 0
0 0 0 0 0 0 0 0
0 0 0 0 0 0 0 0
0 1 0 0 0 0 0 0
```

showing off the distortions together with the informative local features.

The situation can be improved if an integrative algorithm is used instead of this one based on local picture analysis. Let us proceed the former example.

We shall consider the 3 x 3 subsegments of the retina and the following subvectors

$$z'(m,n) = <z_{m-1,n-1}, z_{m-1,n}, z_{m-1,n+1}, z_{m,n-1}, z_{m,n}, z_{m,n+1},$$

(3.22)

$$z_{m+1,n-1}, z_{m+1,n}, z_{m+1,n+1}> .$$

Taking $m = 2,4,6,8,...$, $n = 2,4,6,8,...$ we shall

calculate the weights $w\left[z'(m,n)\right]$ and the following matrix will
be filled in:

$$w\left[z'(2,2)\right] \quad w\left[z'(2,4)\right] \quad \ldots$$
$$w\left[z'(4,2)\right] \quad w\left[z'(4,4)\right] \quad \ldots$$
$$\vdots \qquad \qquad \vdots \qquad \ddots$$

The noisy version of the picture given in the ex
ample gives us then the matrix:

$$\begin{array}{ccccc} 0 & 5 & 5 & 1 & . \\ 2 & 6 & 3 & 2 & . \\ 4 & 6 & 2 & 1 & . \\ 1 & 3 & 1 & 1 & .. \end{array}$$

Let us shortly denote the components of the last
matrix by $w_{i,j}, i=1,2,3,\ldots, j=1,2,3,\ldots$. Once more the 3×3 components
subvectors will be considered and the following relations will
be defined:

$R^I(i,j)$ is satisfied by all the subvectors such that

$$w_{i-1,j-1} + w_{i,j} + w_{i+1,j+1} \;>\; w_{i-1,j} + w_{i-1,j+1} + w_{i,j+1} \qquad (3.23a)$$

and

$$w_{i-1,j-1} + w_{i,j} + w_{i+1,j+1} \;>\; w_{i,j-1} + w_{i-1,j+1} + w_{i+1,j} \;; \qquad (3.23b)$$

$R^{II}(i,j)$ is satisfied by all the subvectors such that

$$w_{i-1,j} + w_{i,j} + w_{i+1,j} \;>\; w_{i-1,j-1} + w_{i,j-1} + w_{i+1,j-1} \;, \qquad (3.24a)$$

and

$$w_{i-1,j} + w_{i,j} + w_{i+1,j} \;>\; w_{i-1,j+1} + w_{i,j+1} + w_{i+1,j+1} \;; \qquad (3.24b)$$

$R^{III}(i,j)$ is satisfied by all the subvectors such that

(3.25a) $w_{i-1,j+1} + w_{i,j} + w_{i+1,j-1} > w_{i-1,j-1} + w_{i-1,j} + w_{i,j-1}$,

and

(3.25b) $w_{i-1,j+1} + w_{i,j} + w_{i+1,j-1} > w_{i,j+1} + w_{i+1,j} + w_{i+1,j+1}$;

$R^{IV}(i,j)$ is satisfied by all the subvectors such that

(3.26a) $w_{i,j-1} + w_{i,j} + w_{i,j+1} > w_{i-1,j-1} + w_{i-1,j} + w_{i-1,j+1}$,

and

(3.26b) $w_{i,j-1} + w_{i,j} + w_{i,j+1} > w_{i+1,j-1} + w_{i+1,j} + w_{i+1,j+1}$.

The relations R^{I}, R^{II}, R^{III} and R^{IV} can be called correspondingly the "North-West-South-East", the "North-South", the "North-East-South-West" and the "West-East" directions of a line going through the centre (i,j) of a segment. Applied to the w_{ij}-–matrix they give the result:

$$R^{II}(2,2) \;—$$

$$\left.\begin{array}{l} R^{II}(2,4) \\ R^{IV}(2,4) \end{array}\right\} \;—$$

indicating the fact that a line of a "North-South" direction is going through the left-half of the original retina-segment (the line is bended horizontally in the lower subsegment).

Some other interesting and effective algorithms of noisy and fuzzy pictures processing at the first levels of structural treatening have been proposed by R. Narasimhan in [24, 25, 26, 27] .

4. Recognition of integral features.

The picture description obtained after the first levels of signal processing has a form of a set of addressed local features. For example, an input signal having the form

```
0 0 0 0 0 0 0 0 0 0
0 0 0 1 1 0 0 0 0 0
0 0 0 1 1 1 0 0 0 0
0 0 0 1 1 1 1 0 0 0
0 0 1 1 0 1 1 1 0 0
0 0 1 1 0 0 1 1 0 0
0 0 1 1 1 1 1 1 1 0
0 0 1 1 1 0 1 1 0 0
0 0 1 1 0 0 1 1 0 0
0 0 0 0 0 0 0 0 0 0
```

may be coded into the form

$$. \;.\;.\;.\;.\;.\;.\;.\;.\;.\;.$$
$$.\qquad\qquad R^{ii}\qquad\qquad .$$
$$.\quad R_1^{iii}\qquad R_2^{iii}\quad .$$
$$.\quad R_1^{i}\qquad\quad R_2^{i}\quad .$$
$$. \;.\;.\;.\;.\;.\;.\;.\;.\;.\;.\;.$$

where the R^{i}, R^{ii} and R^{iii} are the symbols of the following local relations being detected:

R^{i} — "line—end",

R^{ii} — "line—bend",

R^{iii} — "line—branching".

In order to classify the initial picture as a pat_tern the higher—order relationships must be introduced. First of all, the following "mutual position" relations will be necessary:
a) "$R^{'}$ over $R^{''}$"can be defined as a disjunction of all possible cartesian products of two non—empty local relations $R^{'}$, $R^{''}$ such that $R^{'}$ and $R^{''}$ are defined on the corresponding subfamilies of sets $\langle U \rangle^{'}$, $\langle U \rangle^{''} \subset \langle U \rangle$ assigned to the retina—subsegments being local_ized the first over the second one. The word "over" may be inter_preted in the sense of the weight—centres position and in the

upper and lower bounds position as well.

b) "R' on the left of R''" can be defined in similar way, as a disjunction of all possible cartesian products of tow non-empty local relations R', R'' such that R' and R'' are defined on the corresponding subfamilies of sets $\langle U \rangle'$, $\langle U \rangle'' \subset \langle U \rangle$ assigned to the retina subsegments being localized the first on the left to the second one.

Now, the pattern description may be given in the form:

$$R_A = (R_1^{iii} \text{ over } R_1^{v}) * (R_2^{iii} \text{ over } R_2^{i}) * (R^{ii} \text{ over } R_1^{iii}) * (R^{ii} \text{ over } R_2^{iii}) *$$

$$* (R_1^{iii} \text{ on the left of } R_2^{iii}) * (R_1^{i} \text{ on the left of } R_2^{i}).$$

This method of pattern description is invariant with respect to the shiftings, rotations and form changing within reasonable limits. The "letter A" just formally described is legible in spite the noise influence and the print-defects, if the description is sufficient and the local relation tests are strong enough.

The set of mutual position indicating relations can be completed, if necessary, by the relations describing the topological or geometrical structures.

Let us consider an algebra of relations

(4.1) $G = \langle \langle U \rangle, Q_{\langle U \rangle}, \theta, *, ^- \rangle .$

Any relation R belonging to this algebra can de

fine a pattern in the case, if $<\!U\!>$ is interpreted as a family of sets of input signal values. However, if there is a given subset $\{R_i\}$ of some "local" relations, it can be defined a subset $B_{\{R_i\}} \subset Q_{<U>}$ of all relations, which can be expressed as some algebrical com binations of the relations belonging to $B_{\{R_i\}}$. Any relation $R \in B_{\{R_i\}}$ can be expressed by a term composed of the local rela- tions and the symbols of the relation complement⁻, convolution $*$ and cartesian product $\times$. It is clear now, that for a given set of patterns represented by the relations $R_\Lambda, \ldots, R_K$ the set $B_{\{R_i\}}$ must be chosen in such a way that

$$R_\Lambda, \ldots, R_K \in B_{\{R_i\}} . \qquad (4.2)$$

On the other hand, if the last condition holds, the problem arises of constructing the shortest expressions describing the given relations. Owing to the isomorphism between the relations algebra G and the Boolean algebras it is possible to solve this problem using the well known methods of minimiza- tion of the forms of the Boolean functions.

Let us define more exactly, what an "algebraic expression formed by the relations belonging to $\{R_i\}$" does mean:

a) any   symbol R_j of a relation belonging to $\{R_i\}$ is an ex- pression;

b) if q is an expression, then $\bar{q}$ is also an expression;

c) if q', q'' are some expressions, then $q' * q''$ and $q' \times q''$ are also expressions;

d) if $q_1, q_2, ..., q_i, ...$ are some expressions, then $*q_i$ is an expression equivalent to $q_1 * q_2 * ... * q_i * ...$ and $\underset{i}{\times} q_i$ is an expression equivalent to $q_1 \times q_2 \times ... \times q_i \times ...$.

The principles a) – d) are not but some grammatical rules of a formal language describing the relations belonging to $B_{\{R_i\}}$. The set $B_{\{R_i\}}$ is a semantic field of the language, while the given relations $R_A, ..., R_K$ satisfying to (4.2) form a pragmatic field.

Two expressions q', q'' will be called mutually equivalent in formal sense if any one can be transformed into each other using the identity formulas (2.24), (2.27). The problem of proving the formal identity of expressions plays an important role in the structural pattern recognition technique. It is clear that a given pattern can be formally described in several ways and it is necessary to prove the fact that a given expression obtained from a recognitive experiment describes the same pattern as defined by a theoretically deducted expression. The problem of proving the formal identity of the expressions resembles this one of automatic proving the theorems and can be solved using the same tehnique.

The formal identity of expressions should not be confused with the semantical one. The expressions will be called semantically identical if they symbolize the identical relations, that is the relations satisfied by the same sets of realizations. The class of expressions formally identical to an expression q

will be denoted by $h(q)$, while the class of the expressions iden
tical to q in semantic sense by $H(q)$. It is clear that

$$h(q) \subset H(q) . \qquad (4.3)$$

As an example let us take into account a problem
of printed letters recognition. The set of basic relations $\{R_i\}$
may contain the above mentioned relations R^i, R^{ii}, R^{iii} as well as the
higher order relations like "over", "on the left" etc. It is also
possible to define the set of basic relations in another way, in
cluding the relations like "a vertical line-segment", "a horizon
tal line-segment", "a convex arc" etc. The expressions describing
the pattern will not be formally equivalent in both cases, however, they may be equivalent in semantical sense.

A set $<U>$ with a family $\{R_i\}$ of relations describ
ed on it forms a structure in the wide algebraic sense. A pattern
recognition problem is then formally described by a wide algebra
ic sense structure. The pattern recognition problems from the
relation theory point of view can be classified as the construc
tive problems in the theory of relations.

Let us consider two linearly ordered families of
sets $<U>, <V>$ and let R_U , R_V be two relations described on $<U>$
and on $<V>$ correspondingly. Let us denote by Ψ a one-to-one pro
jection of $<U>$ into $<V>$. Let $\{\varphi_i(x)\}$ be a family of one-to-one
projections of the form

(4.4) $\varphi_i : U_i \longrightarrow V_i, \qquad V_i = \Psi(U_i)$.

Otherwise speaking, $\varphi_i(x)$ is a function projecting U_i into V_i if U_i and V_i correspond to each other in the Ψ – projection sense. It is evident, that the set families $<U>$ and $<V>$ as well as the corresponding sets U_i , V_i are pairwise of the same powers.

Let us suppose that the set families $<U>$ and $<V>$ are linearly ordered. We will say that the one-to-one projection Ψ preserves the orders if for any $U_1 \prec U_2 \prec \ldots \prec U_k \in <U>$ (the order being taken in the $<U>$ family sense).

(4.5) $\Psi(U_1) \prec \Psi(U_2) \prec \ldots \prec \Psi(U_k)$

(the order being taken in the $<V>$ -family sense). We will be interested only in the order-preserving Ψ -projections. Any realization $z \in R_U$ is projected by the family of projections $\varphi_i(x)$ into the V -space of sequences; all the sequences so obtained can be considered as a new relation denoted by R_U' . We will say that the relations R_U , R_V are mutually isomorphic if the projections $\Psi , \{\varphi_i(x)\}$ can be chosen in such a way that Ψ is order-preserving and the relation R_U' is identical to R_V .

The following example will illustrate the last concept. Let us take into account the following subfamilies of a set-family describing the state of a retina:

$$<U_{k-1,\ell-1} \quad U_{k-1,\ell} \quad \cdots \qquad \cdot \qquad \cdot$$

$$U_{k,\ell-1} \quad U_{k,\ell}> \quad \cdots \qquad \cdot \qquad \cdot$$

$$\vdots \qquad \vdots \qquad \ddots \qquad \vdots \qquad \vdots$$

$$\cdot \qquad \cdot \qquad \cdots \quad <U_{m-1,n-1} \quad U_{m-1,n}$$

$$\cdot \qquad \cdot \qquad \cdots \quad U_{m,n-1} \quad U_{m,n}> \; .$$

Let us take into account the family of sets $<U_{k-1,\ell-1}, U_{k-1,\ell}, U_{k,\ell-1}, U_{k,\ell}>$ and the following relation describ̲ed on it:

$$R' = \{<10 \; 10>, <01 \; 01>\} \; .$$

Then, let us consider the family of sets $<U_{m-1,n}, U_{m,n}, U_{m-1,n-1}, U_{m,n-1}>$ and the relation

$$R'' = \{<11 \; 00>, <00 \; 11>\} \; .$$

The relations R' and R'' are not identical, of course. Nevertheless, they are mutually isomorphic. Let us define the projection:

$$\Psi(U_{k-1,\ell-1}) = U_{m-1,n} \; , \quad \Psi(U_{k,\ell-1}) = U_{m-1,n-1} \; ,$$

$$\Psi(U_{k-1,1}) = U_{m,n} \; , \quad \Psi(U_{k,\ell}) = U_{m,n-1} \; .$$

It is clear that it is an order–preserving projec̲tion. Then, the following four projections can be defined:

$$\varphi_1(1) = \varphi_2(1) = \varphi_3(1) = \varphi_4(1) = 1, \qquad \text{a)}$$

$$\varphi_1(0) \ = \ \varphi_2(0) \ = \ \varphi_3(0) \ = \ \varphi_4(0) \ = \ 0 \ .$$

The relation R' projected into the family of sets $\langle U_{m-1,n}, U_{m,n}, U_{m-1,n-1}, U_{m,n-1} \rangle$ gives us

$$R''' \ = \ \big\{ \langle 11 \ 00 \rangle, \langle 00 \ 11 \rangle \big\}$$

which is identical with the relation R''. The relation R' and R'' are then mutually isomorphic. This kind of an isomorphism may be called a "shiftening".

An isomorphism between the relations defined on the same family of sets (ordered in different ways) will be called an automorphism. The relation R' in the last example and a relation R^{iv} described on the set family $\langle U_{k,\ell-1}, U_{k-1,\ell-1}, U_{k,\ell}, U_{k-1,\ell} \rangle$ as

$$R^{iv} \ = \ \big\{ \langle 11 \ 00 \rangle, \langle 00 \ 11 \rangle \big\}$$

are automorphic to each other: the first relation describes a horizontal line-segment, while the second one describes a vertical line-segment. This kind of automorphism may be called a "rotation".

It is also possible to define other kinds of isomorphisms like scale-changing etc. On the other hand, let us remark that an optical transformation given by the formula (3.14) is not an isomorphism of relations in the above-mentioned sense.

Let us consider a more detailed general situation.

There are given two families of sets $<U>$, $<V>$ with the relations R_U and R_V, correspondingly. Let us suppose that Ψ is an one-to-one projection of $<U>$ into $<V>$. However, it will be supposed that the projections ψ_i given by (4.4) are unique but not reciprocal. Any realizations of the relation R_U is then supposed to be projected into one and only one realization of R_V and any realization of the relation R_V is supposed to be a projection of at least one realization of R_U. In this case R_V will be called homomorphic to R_U. The following example will illustrate the concept of the homomorphism of the relations.

Let $U, i = 1, 2, ..., 9$ be some trinary sets, consisting of the elements $0, 1, 2$ and let $V_j, j = 1, 2, ..., 9$ be some binary sets consisting of the elements $0, 1$. The set families will be arrayed, as usually, in a square form:

$$<U_1 \quad U_2 \quad U_3 \qquad\qquad <V_1 \quad V_2 \quad V_3$$
$$U_4 \quad U_5 \quad U_6 \qquad\qquad V_4 \quad V_5 \quad V_6$$
$$U_7 \quad U_8 \quad U_9> \qquad\qquad V_7 \quad V_8 \quad V_9> \, .$$

Let us define a relation called "a cross":

$$R_U = \{<101 \ 010 \ 101>, <212 \ 121 \ 212>, <202 \ 020 \ 202>,$$
$$<010 \ 111 \ 010>, <121 \ 222 \ 121>, <020 \ 222 \ 020>\}.$$

The relation

$$R_V = \{<101 \ 010 \ 101>, <010 \ 111 \ 010>\}$$

described on the set family $\langle V \rangle$ can be considered as homomorphic with respect to R_U . Signal levels quantization, reduction or limitation leads, usually, to a homomorphism of relations. However, the signal-reduction technique demonstrated in the former paragraph, based on the averaging technique, cannot be classified as leading to homomorphic transformation in the above-discussed sense.

        Some other kinds of morphisms are also of interest. It may happen that there is given a projection R_U' of R_U into a certain subfamily of sets, and a relation R_V . It will be called that R_V is subisomorphic with respect to R_U if it is isomorphic (respectively, homomorphic) with respect to R_U' .

        On the other hand, a relation R_V will be called overisomorphic (overhomomorphic) with respect to R_U is there exists a projection R_V' of R_V isomorphic (respectively,homomorphic) with respect to R_U . Any projection of a relation is subisomorphic (more exactly, subautomorphic) with respect to the original relation. There is also possible a sub-, over- or isomorphism as well as a sub-, over, or homomorphism between some subrelations. It will be used the term of a partial morphism between R_U and R_V in case there exist some subrelations $R_U' \subset R_U$, $R_V' \subset R_V$ and a kind of morphism between them is stated, which cannot be spread out on the relations R_U, R_V .

        The above given concepts of relations morphisms will be applied to a classification of morphisms existing be-

tween the wide-sense algebraic structures. Let us take into account two algebraic structures

$$T_U = <<U>, \{R_U^i\}> , \qquad\qquad (4.6a)$$

$$T_V = <<V>, \{R_V^i\}> . \qquad\qquad (4.6b)$$

The structures T_U, T_V will be called mutually
isomorphic if there is a set of projections Ψ, $\{\varphi_i\}$ such that
the relations $R_U^i \in \{R_U^i\}$, $R_V^i \in \{R_V^i\}$ are simultaneously pairwise isomorphic. A homomorphism of the structures as well as a sub-,
over- or partial morphism can be defined in similar way. It is
evident that the pattern recognition problems are not necessary
to be considered but within the classes of isomorphic structures.
However, it is possible that not only the morphisms of relations,
but also the morphisms of structures may be of partial type.
For example, the pattern recognition problems connected with
automatic reading English and Russian alphabets can be formally
represented by partially isomorphic structures, so as the relations describing the printed letters "A", "B", "C", "E", "H" and
so on are common in both cases. However, the problem of similari
ty of structures is much deeper. It is well known, for example,
that any practical pattern recognition problem can be effective
ly solved by a lot of formally different ways. On the other hand,
formal similarity of some approaches does not guarantee similar
effectiveness. Therefore, it is necessary to operate with some
more flexible concepts of structural similarity or equivalence

than this one based on the idea of morphism. This cab be reached,
in general, by introducing some distance characteristics into
the space of structure.

Let us suppose, that there is a correspondence
at the lowest level of the structure being compared. It means,
that the set-families $<U>$, $<V>$ and a class of projections Ψ,
$\{\varphi_i\}$ are given. The algebra of relations makes it possible to
describe the meta-relations between the relations $\{R_U^i\}$ as well
as those between the relations $\{R_V^i\}$. The meta-relations can be
expressed in terms of subrelations, convolutions, cartesian prod
ucts etc. Let us remark that the following logical meta-relations
are true: if $A(R)$ denotes that there is given a realization sat
isfying to the relation R, then

a) $A(R) \Longleftrightarrow A(R)$;

b) $A(R) \Longrightarrow A(R_{<U>})$ for any subfamily of sets $<U>$;

c) $A(R') \Longrightarrow A(R)$ for any $R' \subset R$;

d) $A(R') \Longrightarrow A(R' \times R'')$ for any R'' ;

e) $A(R' * R'') \Longrightarrow A(R'), A(R'')$;

f) $A(\bar{R}) \Longrightarrow \neg A(R)$.

These properties make it possible to introduce a
semi-ordering into the set of relations. It will be assumed that

(4.8) $R' \prec R''$ if and only if $A(R') \Longrightarrow A(R'')$.

It can be easily proved that this new meta–relation is reversible (see 4.7a) antisymmetric (if $R' \prec R''$ and $R'' \prec R'$ then $R' \equiv R''$) and transitive. It can be represented by a graph in which the nodes correspond to the relations of the structure un_der consideration and the arrows indicate the logical implications. The following example will illustrate this.

There is given a family of sets:

$$<U_1 \; U_2 \; U_3$$
$$U_4 \; U_5 \; U_6$$
$$U_7 \; U_8 \; U_9>$$

and the following relations:

$$R_1 = \{<100 \; 100 \; 100>,<010 \; 010 \; 010>,<001 \; 001 \; 001>\},$$

$$R_2 = \{<111 \; 000 \; 000>,<000 \; 111 \; 000>,<000 \; 000 \; 111>\}.$$

Let us take into account

$$R_3 = R_1 U R_2 \equiv R_1 \times R_2 = \{<100 \, 100 \, 100>,<010 \; 010 \; 010>,<001 \; 001 \; 001>,$$
$$<111 \; 000 \; 000>,<000 \; 111 \; 000>,<000 \; 000 \; 111>\}$$

and the projection

$$R_4 = R_{3_{<U_1,U_4,U_7>}} = \{<000>,<111>,<100>,<010>,<001>\}.$$

The set of relations $\{R_1,R_2,R_3,R_4\}$ can be semi-ordered in the following way:

$$R_3 \succ R_1, \quad R_3 \succ R_2, \quad R_3 \prec R_4$$

and the graph illustrating the meta-relations will have the fol
lowing form:

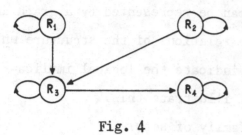

The structure can be compared
from the point of view of simi
larity of the graphs describing
their internal logical micro-

Fig. 4

structure. This general concept
can be improved by additional indexing of the nodes: the nodes
corresponding to the relations mutually isomorphic may have assigned the same index of the class of isomorphic relations. In
the above-given example the same additional index should be attached to the nodes corresponding to the relations R_1 and R_2.

5. Formal languages for image processing.

In the last years a growing attention was paid to
the concept of pattern recognition systems based on the princi-
ple of man-computer interaction. The Illiac III Computing System
designed and built up in the University of Illinois (USA) is the
most represenative example of practical implementation of this
idea. [21, 22, 28, 30] . However, it seems probable that minor
and more specialized pattern recognition and image-processing
systems based on the mini-computers of the third generation will
play an important role in practice. In engineering design, exper

imental data processing, criminological inquiries etc. a possi
bility of control of the decision making process at least at
its crucial stages seems very attractive. The SHOW-and-TELL In
teractive Programming System for Image Processing worked out at
the Department of Computer Science University of Illinois is
based on a formal language for programming the image processing
algorithms. The general idea of this language (described by R.
Narasimhan in [26]) has been also implemented,with some modifi
cations, on a Polish computer ODRA 1204; the language called
PICTURE ALGOL 1204 A should be considered as a versatile tool
for writing down and testing new pattern recognition and image
processing algorithms rather than as a final form of the computer programs destinated for a continuous using. The PICTURE
ALGOL 1204 A makes it possible to operate with binary ("white-
black") graphical pointed information given in the form of a ma
trix of the dimensions $M \times N$, where the number N of columns,
at a given state of matter, is limited by the standard word-size
of the computer ODRA 1204 (24 bits). The number M of rows is
not limited but by the capacity of the operative-memory. Being
realized as a sublanguage imbedded in the ALGOL the PICTURE ALGOL
1204 A has the opportunity of possible realization of its stand
ard procedures on the INTEGER-type variables instead of the REAL
ones. The INTEGER variables M, N as well as some auxiliary vari
ables (BOUND and THR) are declared while a PICTURE ALGOL procedure is introduced.

There are several kinds of standard procedures in
the PICTURE ALGOL. The set algebra procedures make possible to
perform the basic set-algebra operations on the sets of units
the pictures consist of. This corresponds to the basic operations
of disjunction, conjunction and taking a difference of the rela
tions described on binary sets and represented by single realiza
tions. Two other procedures make it possible to erase a picture
and to fill the place (attached to a given arrayed variable) with
zeros or to write there down another picture. Other procedures
of this first group assign the value "true" to a BOOLEAN variable
in the case if a picture consists of zeros only as well as in the
case if two given pictures are identical.

Another group of standard procedures gives the possibility of proving if some point belongs to a given picture (if
the picture is "black" at this point) as well as to add this
point to the picture or to take it away.

The next group of standard procedures makes it
possible to transform the pictures in some more complicated ways.
In general, the way of processing a picture depends on another
picture $Z2$ called "a context" and the result is given in the
form of a third picture $Z3$. To any point z there are assigned
nine points forming its environment (including the point z itself). The addresses of these points are indicated by an auxiliary variable called "direction" taking the values according to
the schema:

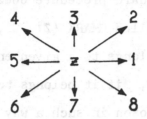

and the address of the point **z** itself is represented by direction $= 0$.

The procedure MARK/Z1, Z3, direction/sets into Z3 all the points belonging to the enviroment at given directions of the points belonging to Z1. For example, if direction $= \,'123'$, that means that the points at the directions 1, 2 and 3 belong to the enviroment of a given point, and if the picture Z1 has the following form

```
0 0 0 0 0 0 0 0
0 0 0 1 1 1 0 0
0 0 0 1 1 1 0 0
0 0 0 1 1 1 0 0
0 0 1 1 1 0 0 0
0 1 1 1 0 0 0 0
0 1 1 1 0 0 0 0
0 0 0 0 0 0 0 0
```

then the picture Z3 obtained from MARK (Z1, Z3, '123') will have the form

```
0 0 0 1 1 1 1 0
0 0 0 1 1 1 1 0
0 0 0 1 1 1 1 0
0 0 1 1 1 1 0 0
0 1 1 1 1 0 0 0
0 1 1 1 1 0 0 0
0 0 1 1 1 0 0 0
0 0 0 0 0 0 0 0
```

This standard procedure does not depend on the context. The next one called CMARK (Z1, Z2, Z3, direction) sets a point into Z3 if it belongs to a given environment of a point of Z1 and simultaneously, if it belongs to a given picture Z2. The picture Z2 can be chosen in such a way that Z3 will not be written down but inside a given segment. For example, if we put in the former example Z2 in the form

```
0 0 0 0 0 0 0 0
0 0 0 0 0 0 0 0
0 0 1 1 1 1 0 0
0 0 1 1 1 1 0 0
0 0 1 1 1 1 0 0
0 0 1 1 1 1 0 0
0 0 0 0 0 0 0 0
0 0 0 0 0 0 0 0
```

the standard procedure CMARK (Z1, Z2, Z3, '123') will give the result

```
0 0 0 0 0 0 0 0
0 0 0 0 0 0 0 0
0 0 0 1 1 1 0 0
0 0 1 1 1 1 0 0
0 0 1 1 1 0 0 0
0 0 1 1 1 0 0 0
0 0 0 0 0 0 0 0
0 0 0 0 0 0 0 0
```

The procedures MARK and CMARK can be generalized in a following way. One introduces a BOOLEAN variable bex(THR), where the INTEGER called THR denotes the weight of the entourage of a given point of Z2 at the indicated directions. The value of bex is equal "true" if the weight computed for the entourage of a current point of Z2 is more or equal to a threshold value. In particular, taking the procedure THRESHOLDCMARK (Z1, Z2, Z3, '12345678', THR<8) for a picture Z1 of the form

```
0 0 0 0 0 0 0 0 0
0 0 0 0 1 1 0 0 0
0 0 0 1 1 1 0 0 0
0 0 1 1 1 1 0 0 0
0 0 0 1 1 1 0 0 0
0 0 0 1 1 1 0 0 0
0 0 0 1 1 1 1 0 0
0 0 1 1 1 1 1 0 0
0 0 0 0 0 0 0 0 0
```

we obtain the contour of the image:

```
0 0 0 0 0 0 0 0 0
0 0 0 0 1 1 0 0 0
0 0 0 1 1 1 0 0 0
0 0 1 1 0 1 0 0 0
0 0 0 1 0 1 0 0 0
0 0 0 1 0 1 0 0 0
0 0 0 1 0 1 1 0 0
0 0 1 1 1 1 1 0 0
0 0 0 0 0 0 0 0 0
```

Some further possibilities are connected with the concept of a "chain". The chain consists of all points belonging both to the pictures $Z1$ and $Z2$ at a given direction apart from a point belonging to $Z1$ up to the moment when the entourage can not be extended in this direction. This can be illustrated by the following example.

If there are given the pictures $Z1$ and $Z2$, correspondingly, in the forms:

```
0 0 0 0 0 0 0 0          0 0 0 0 0 0 0 0
0 0 0 0 0 0 0 0          0 0 1 1 1 1 0 0
0 0 0 0 0 0 0 0          0 1 1 1 1 1 1 0
0 0 1 0 0 0 0 0          0 1 1 1 1 1 1 0
0 0 0 1 0 0 0 0          0 1 1 1 1 1 1 0
0 0 0 0 0 0 0 0          0 1 1 1 1 1 1 0
0 0 0 0 0 1 0 0          0 0 1 1 1 1 0 0
0 0 0 0 0 0 0 0          0 0 0 0 0 0 0 0
```

a procedure called CHAIN ($Z1$, $Z2$, $Z3$, direction) for direction '1,2' will give the picture $Z3$ in the form

```
0 0 0 0 0 0 0 0
0 0 0 0 1 0 0 0
0 0 0 1 0 1 0 0
0 0 1 1 1 1 1 0
0 0 0 1 1 1 1 0
0 0 0 0 0 0 1 0
0 0 0 0 0 1 0 0
0 0 0 0 0 0 0 0
```

The procedure CHAIN combined with a threshold procedure makes it possible to detect some local features of fuzzy pictures. This possibility is offered by a standard proce_ dure called THRESHOLDCHAIN (Z1, Z2, Z3, direction, bex). The procedure plots the chain of Z1 with respect to Z2 and writes it down into Z3 only in the case, if the BOOLEAN variable bex is equal "true"; this last variable is defined by the inequality

$$THR > w$$

where THR is an INTEGER-type variable denoting the weight of the chain and w is a given threshold value.

Let us take the following example. There is given a disturbed picture Z1:

```
0 0 0 1 0 1 0 1
1 0 0 1 1 0 0 1
0 0 1 0 0 1 1 0
0 0 1 1 1 0 0 1
1 1 0 0 0 1 0 0
0 1 1 0 1 1 0 0
1 1 1 0 0 0 1 1
0 1 1 0 1 0 0 0
```

A hypothesis that it describes a fuzzy line having the direction '26' against this one that the direction is '48' is to be proved. Two standards Z21, Z22 having the same weight equal 22 will be introduced:

```
1 1 0 0 0 0 0 0          0 0 0 0 0 0 1 1
1 1 1 0 0 0 0 0          0 0 0 0 0 1 1 1
0 1 1 1 0 0 0 0          0 0 0 0 1 1 1 0
0 0 1 1 1 0 0 0          0 0 0 1 1 1 0 0
0 0 0 1 1 0 0 0          0 0 1 1 1 0 0 0
0 0 0 0 1 1 1 0          0 1 1 1 0 0 0 0
0 0 0 0 0 1 1 1          1 1 1 0 0 0 0 0
0 0 0 0 0 0 1 1          1 1 0 0 0 0 0 0
```

Now, let us perform the two procedures: THRESHOLD
CHAIN ($Z1$, $Z21$, $Z31$, '026', THR > 1) and THRESHOLDCHAIN ($Z1$, $Z22$,
$Z32$ '048', THR > 0). As a result we obtain the pictures $Z31$ and
$Z32$:

```
0 1 0 0 0 0 0 0          0 0 0 0 0 0 1 1
1 0 0 0 0 0 0 0          0 0 0 0 0 1 0 1
0 0 1 1 0 0 0 0          0 0 0 0 0 1 1 0
0 0 1 1 1 0 0 0          0 0 0 1 1 0 0 0
0 0 0 1 0 1 0 0          0 0 0 0 1 0 0 0
0 0 0 0 1 1 0 0          0 1 0 0 0 0 0 0
0 0 0 0 0 0 1 0          1 1 1 0 0 0 0 0
0 0 0 0 0 0 0 0          0 1 0 0 0 0 0 0
```

The second one, having greater weight, indicates
the more probable direction of the line. Now, taking the next
procedure CHAIN ($Z32$, $Z22$, '26') we obtain a regenerated picture:

```
0 0 0 0 0 0 1 1
0 0 0 0 0 1 1 1
0 0 0 0 1 1 1 0
0 0 0 1 1 1 0 0
0 0 1 1 1 0 0 0
0 1 1 1 0 0 0 0
1 1 1 0 0 0 0 0
1 1 0 0 0 0 0 0
```

It may be narrowed, if necessary, by taking Z7 from CHAIN (Z5, Z6, Z7, '0'), for Z6 having the form

```
0 0 0 0 0 0 0 1
0 0 0 0 0 0 1 0
0 0 0 0 0 1 0 0
0 0 0 0 1 0 0 0
0 0 0 1 0 0 0 0
0 0 1 0 0 0 0 0
0 1 0 0 0 0 0 0
1 0 0 0 0 0 0 0
```

The resulting picture Z7 will have the form iden‐tical with Z6.

For the detection of the local relations describ‐ed by several alternative realizations a standard procedure TRANSFORM (Z1, Z2, Z3, bf) has been introduced. The symbol "bf" stands for a string of numbers from 0 up to 8, some of which can be underlined and some sequences of which can be separated by the symbol "+". A notation TRANSFORM ($Z1$, $Z2$, $Z3$, '2̲3̲ + 4̲5̲') indicates, for example, that the picture $Z3$ will consist of all the points belonging to $Z1$ and such that the environment of any point with respect to the picture $Z2$ in the direction '3' is non-empty and in the direction '2' is empty or it is non-empty in the direction '4' and empty in the direction '5'. For example, in order to detect all possible line-ends the following procedure can be used: TRANSFORM ($Z1$, $Z2$, $Z3$, '1̲2̲345678 + 12̲345̲678 + 12̲345̲678 + 12̲345̲678 + 12̲345̲678 + 12̲345̲678 + 12̲345̲678 + 12̲345̲678). For the picture $Z1$ and $Z2$ having the forms

```
0 1 0 0 0 0 0 1        0 0 0 0 0 0 0 0
0 1 0 0 1 1 1 0        0 0 0 0 0 0 0 0
0 0 1 0 1 0 0 0        0 0 0 0 0 0 0 0
0 0 1 1 1 1 1 0        0 0 0 0 0 0 0 0
0 1 1 0 0 0 0 0        1 1 1 1 0 0 0 0
0 0 0 1 1 1 0 0        1 1 1 1 0 0 0 0
0 1 1 1 0 0 1 0        1 1 1 1 0 0 0 0
0 0 0 0 0 0 0 1        1 1 1 1 0 0 0 0
```

we obtain the resulting picture $Z3$:

```
0 0 0 0 0 0 0 0
0 0 0 0 0 0 0 0
0 0 0 0 0 0 0 0
0 0 0 0 0 0 0 0
0 1 0 0 0 0 0 0
0 0 0 0 0 0 0 0
0 1 0 0 0 0 0 0
0 0 0 0 0 0 0 0
```

Some other standard procedures of the PICTURE ALGOL 1204 A are used for the input/output operations.

Several image-processing language imbedded in the FORTRAN programmins language (PAX, PAX II, COMPAX, STANDPAX) are also worthy to be mentioned here. The language PAX II allows processing of pictures of arbitrary sizes and multilevel shading. (19). The progress in this domain will favor the development of pattern recognition methods. However, the problem of formal lan guages in pattern recognition is much wider. Picture is one of the basic forms that the scientific, technical and a lot of other kinds of information can be fixed. Machinery designs, electronic schemas, topographical charts, chemical structural formulas etc. can be considered as some formal languages themselves. The prob lem of pattern recognition is a problem of translation of a "planar" language into a "linear" one, consisting of linear strings of symbols. However, a general theory of planar languages does not exist but in an initial state. It is evident that a planar language can,in general,be defined like a linear one, as an ordered pair (20):

$$L = \langle A, F \rangle \qquad (5.1)$$

where A is a set of basic elements (symbols, words) and F is a set of the "expressions" correct in some formal sense. The main difference consists in the definition of the structure of an "expression". In linear languages, as it was illustrated in the

4th Chapter, this is given by a binary associative and non-com
mutative operation of "concatenation" of the strings of element
belonging to the alphabet A. In planar languages the situation
gets more complicated. The mutual position of two or more frag-
ments of a picture is usually characterized by the qualifica-
tions like: "above" or "below", "near" or "far", "inside" or
"outside" etc. From a formal point of view these kinds of quali
fications are reversible, antisimmetric and transitive. Otherwise
speaking, they define some kinds of semiordering in the space
of expressions containing the set A as its subset. In a particular case, if a single ordering rule is defined and the order
is linear, a simple concatenation is defined and we obtain a
linear language.

In the grammar of planar languages several levels
can be usually specified. At the lowest level the intensity of
shadow and the colours at the plane-points are specified; this
is the analogy to a phonetical or a semiotical level. The level
of basic graphical symbols description corresponds to a morphological level in the structural linguistics. The higher levels
of the theory of planar languages deal with the principles of
graphical symbols combinations; this corresponds to a syntactical
level of a grammar. An investigation of the structural grammars
of the planar languages used in practice could enrich the gener
al theory with experimental facts, which should be taken into
account while the theory is developed. One of the basic theoret

ical problems is this of an equivalence of planar and linear
languages. So as any picture can be transformed with an arbitrary
exactness into a sequence of binary symbols, and, on the other
hand, any binary message can be expressed in a written graphical
form, the potential equivalence of the linear and planar languages in general cannot be called in question. However, the
problem still exists of finding out a linear form for a given
planar language, in some optimum way. The problem of formal description of the forms and dimensions of the machine parts and
of their sets for the engineering design automation is a typical
example of the general problem of equivalence. Till now, the sat
isfactory results in this last field have been obtained in the
domain of formal description of some simpler machinery parts:
flat or rotary ones. The problem-oriented machinery design
languages APT, ADAPT (USA), CLAM, PROFILEDATA, COCOMAT (G.B.),
GEOMETR 66 (USSR) are typical examples of such languages. It is
evident that the requirements of exactness do not allow a discrete point-approximation of a geometrical form like this one
used in the above-mentioned picture-processing languages. The
geometrical languages should be rather based on the geometric-
analytical description of some basic elements: points, straight
lines, planes, triangles, rectangles, circles, spheres etc. in
an euclidean space provided by a system of coordinates. The high
er-order geometrical forms can be obtained from the simpler ones
by the set-algebra operations. A general geometrical language

should contain a new kind of variables, the GEOMETRIC ones, besides the INTEGER, REAL and BOOLEAN ones. The values of the GEOMETRIC variables should be equal "empty" or should be given in the form of a set of geometric -analytical inequalities. For example, a GEOMETRIC called "triangle" can be characterized by the inequalities

$$ax + by + c \; < \; 0$$

$$dx + ey + f \; < \; 0$$

$$gx + hy + i \; < \; 0$$

where $a, b, c, d, e, f, g, h, i$ are some real coefficients satisfying to some additional conditions. It is possible to call a subroutine for proving the fact that a given point z belongs to the "triangle" or not, that two geometrical forms have a nonempty common part etc., so as the question will lead to some numerical calculations, fully algorithmized. However, the geometrical language should offer further possibilities. The necessity of proving some topological relations: of a geometrical object fully included inside another one or being tangent to it, of two or more geometrical objects being situated on the same side of a curve or surface, may also arise. Some geometrical objects of arbitrary forms may be also necessary. Standard geometrical processing: shiftings and rotations of geometrical objects with respect to some system of coordinates or one with respect to another one, generation of new geometrical objects, variation of dimensions till a geometrical, topological or some other condition is fulfilled etc. would provide important facilities for the designer. Existence of such a powerful tool can limit or eliminate the necessity of automatic processing and recognition of some kinds of information given in the planar form. The technological information, for example, can be completely processed in the linearly coded form up to the moment when the result is printed in the form of a standard technological documentation, while the pictoral version of information could be used for the man-surveillance only. Otherwise speaking, the pattern recognition and image processing systems should be considered in the wide context of information processing systems exploiting all the opportunities provided by any form of the information.

List of basic symbols.

U $\quad$ a sum of sets or a disjunction of relations,

∩ $\quad$ an intersection of sets or a conjunction of relations,

∸ $\quad$ a difference of sets or of relations,

∈ $\quad$ inclusion of an element,

⊂ $\quad$ inclusion of a subset or of a subrelation,

x $\quad$ single Cartesian product of sets or relations,

X $\quad$ multiple Cartesian product of sets or relations,

− $\quad$ a complement of a relation,

V $\quad$ a logical disjunction (alternative),

Λ $\quad$ a logical conjunction,

⌐ $\quad$ logical negation,

⟹ $\quad$ logical implication,

⟺ $\quad$ logical equivalence (two-sided implication)

<> $\quad$ linearly ordered set or segment,

{} $\quad$ unordered set,

$\{S\}_i$ $\quad$ i-th set of input messages (i-th pattern)

s $\quad$ realization of an input message,

$\{Z\}$ $\quad$ set of received signals,

z $\quad$ realization of a received signal−set of decisions

$\{Y\}$ $\quad$ set of decisions,

y $\quad$ a decision,

N,M $\quad$ integers indicating the dimensionality of input signals,

R $\quad$ a relation

$R_{<Z>}$ $\quad$ projection of a relation R on the subfamily of sets $<Z>$

$\varkappa$ cardinal number of a set,

U   set of input signal values describing the state of retina,

U_i set of the i-th input signal component values,

$\varrho(z^*,z)$ distance between the vectors z^* and z.

θ empty relation

$Q_{<U>}$ trivial relation described on the set-family $<U>$.

Bibliography

A. General

[1] Aizerman M.A., Braverman E.M. and Rosoneer L.I.:"Teoreti-
    českije osnovy metoda potencjalnych funkcij
    v zadače ob obučenii avtomatov raspredelenju
    vchodnych sitacij na klassy – Theoretical founda
    tions of the potential-functions method in the
    problem of teaching automata, the classification
    of input situations" in Russian. Avtomatika i
    telemechanika, vol. XXV, N° 6, 1964.

[2] Avtomatičeskoje čtenje teksta – Automatic text reading, in
    Russian, VINITI Moscow, 1967.

[3] Braverman E.M.: " O metode potencjalnych funkcij – About
    the potential-function methods", in Russian. Avto
    matika i telemechanika, V. XXVI N° 12, 1965.

[4] Čitajuščije avtomaty i raspoznavanje obrazov – Reading
    automata and recognition of patterns, in Russian
    Naukova Dumka, Kiev, 1965.

[5] Čitajuščije ustrojstva – Reading devices, in Russ. VINITI
    Moscow, 1962.

[6] Fu K.S.: "Learning control systems. Computer and informa-
    tion sciences. Washington 1964.

[7] Kulikowski J.L.:"Cybernetycze uklady rozpoznajace – Cyber-
    netical recognition systems". in Polish, PWN, War
    saw, to be printed.

[8] Opoznaje obrazov – Pattern recognition, in Russian, Izd.
    Nauka, Moscow, 1968.

[9] Principles of self-organization, Trans. of the University
    of Illinois Symp. on Self-Organization, Pergamon
    Press, 1962.

[10] Rasoznavanje obrazov.Konečnyje avtomaty i relejnyje ustro
    jstva – Recognition of patterns. Finite automata
    and switching systems, in Russ. Izd. Nauka,

Moscow, 1967.

[11] Sebestyen G.S.: "Decision making processes in pattern rec̲ognition. The McMillan Company, New York, 1962.

[12] Trudy III Vsesojuznoj konferencji po informacjonno-poisko̲ vym sistemam i avtomatizirovannoj obrabotke naučno-techničeskoj informacji - Transactions of the 3rd all-Union conference on the information-retrieval systems and automatic processing of scientific and technical information, in Russian Vol. III Avtomaticeskije citajuscije ustrojstva. VINITI, Moscow 1967.

B. On algebraic and structural methods

[13] Avtomatičeskij analiz složnych izobraženij - Automatic a̲nalysis of composite patterns, in Russian, Izd. MIR, Moscow, 1969.

[14] Breeding K.J.: "Grammar for a pattern description language. Dept. of Computer Science, Un. of Illinois, Rep. N° 177, May 1965.

[15] Frischkopf L.S., Harmon L.D.: "Machine reading of cursive script. 4th London Symposium of Information Theo̲ry. Butterworth, London, 1961.

[16] Fajn V.S.: "Opoznavanje izobraženij - Recognition of pat-terns", in Russ. Izd. Nauka, Moscow, 1970.

[17] Kulikowski J.L.: "Niektore problemy strukturalnej analizy obrazov zložonych - Some problems of structural analysis of composite patterns", in Polish. Archivum Automatyki i Telemechaniki Vol XV, N° 3 1970.

[18] Kulpa Z., Szydlo Jezyk: "PICTURE ALGOL 1204 A. Opis užyt-kovy - The language PICTURE ALGOL 1204 A. A use manual, in Polish Institute of Automation Polish Academy of Sciences, a report, Warsaw 1971.

[19] Lipkin B.S.(ed.), Rosenfeld A.: " Picture processing and
     psychopictorics", Acad. Press New Y.-Lond. 1970.

[20] Marcus S.: "Algebraic linguistics; analytical models. New
     York-London, 1967.

[21] McCormick B.H.: "The Illinois pattern recognition computer"
     (Illiac III) Digital Computer Lab. Un. of Illinois
     Rep. N° 148, 1963.

[22] McCormick B.H.: "Illiac III computer system. Brief descrip-
     tion and annotated bibliography. Dept. of Comp.
     Science Un. of Illinois, File N° 841, June 1970.

[23] McCormick B.H.and Schwebel J.C.: "Properties of a discrete
     space preserved by image processing relations.
     Dept. of Computer Science Univ. of Illinois File
     N° 769, July 1968.

[24] Narasihman R.: "A linguistic approach to pattern recogni-
     tion. Digital Computer Lab. Un. of Illinois, Rep.
     N° 121, July 1962 (see also   13).

[25] Narasihman R.: " BUBBLE SCAN I program. Digital Computer
     Lab. Univ. of Illinois Rep. N° 167.Aug. 1964
     (see also   13).

[26] Narasihman R.: "Labelling schemata and syntactic descrip-
     tion of pictures. Inform. and Control Vol. 7 151-
     179, 1964 (see also   13).

[27] Narasihman R.: "Syntax-directed interpretation of classes
     of pictures. Commun. ACM Vol 9, N° 3, 1966 (see
     also   13).

[28] Narasihman R., Witsken J.R. and Johnson H.: " BUBBLE TALK
     the structure of a program for on-line conversa-
     tion with Illiac II", Digital Computer Lab. Un.
     of Illinois File N° 604, July 1964.

[29] Polakov V.G.: " O cifrovoj technike sčityvanja i analiza
      konturov – About a digital technique of reading
      and analysis of contours" in Russian, Opoznanje
      obrazov. Teoria peredači informacji. Izd. Nauka
      Moscow, 1965.

[30] Read J.: "SHOW-and-TELL. An interactive programming system
      for image processing. Dept. of Computer Science
      Univ. of Illinois Report N° 429, February 1971.

[31] Rogers J. and Tanimoto T.: "A computer program for classi-
      fying plants. Science, Vol. 132, 1960.

[32] Schwebel J.C.: "Use of graph transformations to character
      ize an image: an illustrative example" Dept. of
      Computer Science Univ. of Illinois, File N° 770
      July 1968.

[33] Sherman H.: "A quasi-topological method for the recogni-
      tion of line patterns, Information Processing"
      UNESCO, Paris, 1960.

[34] Staniszkis M.: "Porownywanie struktur punktowych – Compar
      ison of point structures" in Polish, Institute of
      Automation Polish Academy of Sciences (a Report)
      Warsaw, 1971.

[35] Strukturnoje opoznavanje i avtomatičeskoje čtenje – Struc-
      tural recognition methods and automatic reading
      in Russian, VINITI, Moscow, 1970.

Contents